CARP FARMING CHRONICLES

A Journey Into Freshwater Carp Aquaculture

Dive into the World of Carp Farming and Discover the Thriving Market for Sustainable Freshwater Fish

Dr. Fabian Felicity

Table of Contents

CHAPTER ONE

Introduction

Carp farming, often known as carp aquaculture, is a growing segment of the aquaculture industry, accounting for a major portion of worldwide fish output. Carp, a freshwater fish in the Cyprinidae family, has become a popular choice for fish farming owing to its hardiness, flexibility, and high reproductive rate.

This article seeks to give a full overview of carp farming, including important topics such as carp species selection, carp farm setup, and the necessary aquaculture technologies and infrastructure for effective production.

Understanding Carp Farming

Carp farming is the regulated production of carp in confined habitats, such as ponds or tanks, to supply the growing demand for fish protein.

Carp are known for their omnivorous diet, which allows them to adapt to different feeding regimens and reduces their reliance on costly feeds. Furthermore, their capacity to survive in a variety of environmental situations leads to carp farming's general success throughout geographies.

Recognizing the numerous advantages of carp farming is an important part of comprehending it.

Carp are recognized for their fast development, which allows for short turnaround times in manufacturing. Their capacity to convert a variety of food sources into high-quality protein makes them an efficient and cost-effective aquaculture option.

Furthermore, carp farming creates job possibilities, particularly in rural regions, and helps to ensure food security by providing a consistent source of protein.

Selecting Carp Species

Choosing the appropriate carp species is a vital choice that has a considerable impact on the success of a carp farming operation. Common carp (Cyprinus carpio) is the most widely grown species because of its flexibility, rapid development rate, and tolerance to changing environmental circumstances. Mirror carp, leather carp, and scaled carp are typical variants of the common carp species, each with its unique traits.

In addition to common carp, additional species like grass carp (Ctenopharyngodon idella), silver carp (Hypophthalmichthys molitrix), and bighead carp (Hypophthalmichthys nobilis) are often used in carp farming systems. These species are recognized for their distinct eating patterns, such as filter feeding in the case of silver and bighead carp, which may help to enhance water quality in polyculture systems.

When choosing a carp species, consider market demand, local climatic conditions, and the anticipated production size. Understanding each species' unique needs and habits is critical for establishing a balanced and successful carp farming enterprise.

Setting Up A Carp Farm

The effective creation of a carp farm requires careful planning and consideration of numerous elements. First and foremost, choosing a proper location is critical. Carp farms are often found in places with ample freshwater, ideally free of contaminants. Ponds are a popular alternative for carp farming because they provide a controlled environment for maintaining water quality and enable simple monitoring of fish health.

Pond preparation is an important step after selecting a location. This includes cleansing the pond of trash, leveling the bottom, and assuring appropriate water retention. Adequate aeration systems, such as paddlewheels or air pumps, may be required to maintain ideal oxygen levels, particularly in intensive agricultural operations.

Water quality control is critical in carp farming. Parameters like as dissolved oxygen, pH, and ammonia levels must be tested and monitored regularly to ensure fish health. Proper nutrition is also essential, and farmers must provide a balanced feed that matches the nutritional needs of the selected carp species.

CHAPTER TWO
Aquaculture System And
Infrastructure

The choice of aquaculture technologies and infrastructure has a significant impact on the efficiency and production of a carp farm. Carp farming is divided into three major categories: extensive, semi-intensive, and intense systems, each with its own set of needs and advantages.

Extensive systems entail rearing carp in natural or man-made ponds with little human involvement. This low-input strategy is appropriate for small-scale operations, but it may not optimize production efficiency. Semi-intensive systems achieve a compromise between natural and managed habitats, usually with supplementary feeding and rudimentary water quality management.

In contrast, intensive systems have high stocking densities, improved water recirculation, and intense feeding procedures. While these systems may enhance output levels, they need complex infrastructure, close monitoring, and more expenditure.

Well-designed pond structures, efficient water supply and drainage systems, and secure fencing to avoid predators are all critical infrastructure components. In intensive systems, including technology like biofilters and water recirculation systems improves water quality while lowering environmental effects.

Carp farming is a dynamic and commercially feasible option in the aquaculture industry. Understanding the subtleties of carp farming, from species selection to farm setup and aquaculture system implementation, is critical for success.

As worldwide demand for fish protein rises, carp farming emerges as a dependable and sustainable alternative, generating economic possibilities while also contributing to food security. With the right expertise and careful management, carp farming can be a profitable operation that meets the rising need for efficient and responsible food production.

Water Quality Management

Water quality control is a vital component of effective carp farming since it influences the fish's general health and production. Carp, being a freshwater fish, are particularly sensitive to the condition of the water in which they live.

Proper water quality management ensures that the aquatic environment supports optimum growth and development.

Maintaining appropriate dissolved oxygen levels is an important aspect of water quality management.

Carpets need enough oxygen levels to breathe, and a lack of oxygen may cause stress, lower development rates, and an increased susceptibility to illness. To avoid these complications, oxygen levels must be monitored regularly and supplemented as needed.

Another important consideration is keeping proper pH levels in the water. Carp thrive in slightly alkaline to neutral environments, with pH ranging from 6.5 to 8.5. Fluctuations outside of this range may harm the fish's metabolism and general health.

Regular pH testing and modifications assist in maintaining a stable and optimal environment for carp farming.

Additionally, regulating ammonia and nitrite levels is critical. Ammonia, a byproduct of fish waste, may be harmful at high levels. Nitrite, which is formed during the breakdown of ammonia, is likewise toxic to carp. Implementing effective filtration systems and water exchange procedures helps maintain these factors under control, resulting in a healthy habitat for the fish.

Proactive steps to avoid the buildup of hazardous pollutants, such as organic debris and silt, are also essential. Regular pond and tank cleaning, along with the use of correct aeration devices, helps to preserve water purity and quality. This, in turn, contributes to the general health and expansion of the carp population.

Feeding And Nutrition In Carp Farming

Feeding and nutrition are critical components of effective carp farming operations. Carp are omnivore fish with distinct food needs at various phases of their life cycle. Providing a well-balanced and nutritionally full diet is critical for promoting optimum development, reproduction, and health.

In the early phases of carp farming, high-quality beginning meals are critical. These meals normally include a well-balanced combination of protein, lipids, carbs, vitamins, and minerals to help juvenile carp grow and develop quickly. As the fish age, their food may need to be modified to match their increasing nutritional requirements.

Protein is an essential component of carp diets, particularly in the early phases of development. Fishmeal, soybean meal, and other protein sources are prominent ingredients in carp diet. The protein composition of the meal impacts the fish's muscular development, and low protein levels might cause stunted growth.

In addition to protein, the proper ratio of fats and carbs is required. Fats offer energy, but carbs improve the overall nutritional profile. Monitoring dietary lipid levels is critical for preventing disorders like fatty liver disease, which may result from excessive fat consumption.

Vitamins and minerals are also essential for carp nutrition. These micronutrients promote a variety of physiological tasks, including bone growth, immune system support, and reproduction. To guarantee a well-rounded and comprehensive diet, the fish's nutritional demands must be assessed regularly, and the feed formulation adjusted accordingly.

CHAPTER THREE

Health And Disease Management

Ensuring the health of carp stocks is a primary issue in aquaculture, and good disease control is crucial for avoiding losses and sustaining profitability.

Carps are vulnerable to a variety of illnesses, including bacterial, viral, and parasitic infections. Implementing proactive health management measures is critical for reducing the risks associated with these dangers.

Regular health monitoring, including visual inspections, behavioral observations, and, when required, laboratory testing, is critical in disease prevention. Early diagnosis of possible concerns enables early management and decreases the chance of disease propagation among the community. Quarantine methods for new fish arrivals and strong biosecurity rules help to prevent infections from entering the farm.

Vaccination programs are becoming more frequent in carp farming to guard against certain illnesses. Vaccines activate the fish's immune system, offering resistance to common diseases. A successful disease prevention plan relies on proper vaccination delivery, adherence to specified schedules, and record-keeping.

In situations when illnesses do emerge, timely and precise diagnosis is critical for adopting suitable treatment strategies. To establish a complete and successful disease control strategy, consult with aquatic veterinarians and fish health specialists. Antibiotics, antiparasitic medicines, and other pharmacological treatments are possible treatment choices.

As previously noted, regular water quality management methods play a crucial role in illness prevention. Clean and well-maintained aquatic habitats decrease stress for fish and lower the risk of opportunistic illnesses. In addition, appropriate waste management and the evacuation of infected persons may aid in the containment and control of epidemics.

Breeding & Reproduction

Successful breeding and reproduction are critical components of maintaining a carp farming enterprise. Understanding carp reproductive biology and adopting good breeding procedures are critical to maintaining a constant and sustainable supply of fingerlings for growth-out.

Carps usually attain sexual maturity between two and five years, depending on the species and environmental circumstances. Creating spawning circumstances, such as providing optimal nesting places and maintaining water quality standards, is critical for triggering reproductive activity in adult carp.

Hormone-induced spawning may be used in controlled breeding operations to help synchronize and enhance the reproductive process. Hormones are administered to induce ovulation and spermiation, which facilitates the collecting of eggs and sperm for artificial fertilization.

This technology enables precise control over breeding periods and the selection of particular breeding pairings to improve desired features.

Once fertilized, carp eggs must be carefully incubated. During the incubation phase, water temperature, aeration, and predator protection must all be carefully managed.

These parameters have a direct impact on hatching success, and careful attention to detail is required to achieve a high larval survival rate.

After hatching, the fry goes through several developmental phases before reaching the fingerling stage.

CHAPTER FOUR

Market Trends For Carp Farming

Carp farming has seen changing market patterns influenced by shifting customer choices, environmental concerns, and advances in aquaculture operations. Understanding and reacting to these tendencies is critical to the long-term viability and profitability of carp farming enterprises.

One noticeable trend is the rising desire for responsibly sourced and ecologically friendly seafood. Consumers are becoming more aware of the environmental effects of their food choices, resulting in an increasing desire for goods that use ethical aquaculture techniques. Carp producers may profit from this trend by using sustainable farming practices, reducing environmental impact, and increasing transparency in their production operations.

Another noticeable development is the emergence of value-added carp products on the market. Processed carp goods, such as fillets, smoked carp, and carp-based snacks, provide customers with practical and novel solutions.

Diversifying product offers and experimenting with novel processing processes may boost market competitiveness and adapt to changing customer tastes.

Globalization and international commerce have a significant impact on carp farming market trends. Carp product expansion requires access to new markets as well as the capacity to fulfill international quality and safety requirements.

Keeping up with worldwide market trends, trade rules, and customer tastes across countries is critical for establishing carp farming enterprises in the global marketplace.

Aquaculture technology developments including automation, data analytics, and precision farming are defining the future of carp farming. Integrating this technology into agricultural processes may increase efficiency, lower operating costs, and improve overall farm management. Farmers who embrace innovation are better prepared to face the difficulties of a dynamic and competitive market.

Finally, successful carp farming requires effective control of water quality, feeding and nutrition, health and disease management, breeding and reproduction, and market trends. Adopting best practices in these areas assures the long-term viability and profitability of carp farming operations, therefore helping industry development and fulfilling the changing demands of global customers.

Sustainable Practices In Carp Aquaculture

Carp aquaculture is critical to satisfying the worldwide need for fish protein. However, the sector confronts issues with environmental impact, resource usage, and economic viability. Implementing sustainable procedures is critical to ensuring the long-term profitability of carp farming. This section delves into the fundamental components of sustainable carp aquaculture operations, as well as the issues they provide.

1. Environmental Impact and Resource Utilization.

Carp farming may have serious environmental repercussions if not handled properly. Concerns include the discharge of surplus nutrients into water bodies, the abuse of antibiotics, and the escape of farmed fish into the wild. To overcome these difficulties, farmers are using strategies like integrated multi-trophic aquaculture (IMTA), which includes raising various species simultaneously to produce a balanced environment. IMTA not only decreases environmental effects but also improves resource use by converting waste from one species into nutrients for another.

Efforts are also underway to reduce antibiotic usage in carp farming. The development of alternative disease management measures, like as probiotics and vaccinations, helps to promote sustainable practices by lowering the danger of antibiotic resistance and minimizing the effect on aquatic environments.

2. Water Management and Efficiency.

Sustainable water management is crucial in carp aquaculture for maintaining water quality and conserving this valuable resource. Recirculating aquaculture systems (RAS) are gaining popularity because they reuse water, reduce environmental effects, and ensure efficient resource usage. RAS entails filtering and purifying water for reuse, lowering the requirement for continuous freshwater input, and minimizing the release of nutrient-rich effluents into natural water bodies.

Furthermore, using water conservation methods, such as correct pond design and maintenance, contributes to maintaining ideal circumstances for carp development while reducing the environmental impact of aquaculture operations.

CHAPTER FIVE

Challenges And Solutions

1. Disease Management.

Disease outbreaks are a severe difficulty for carp aquaculture, resulting in economic losses and environmental difficulties. Traditional disease management approaches often entail the use of chemicals, which contribute to water contamination and antibiotic resistance.

Sustainable solutions prioritize preventative measures such as biosecurity standards, optimal nutrition, and selective breeding to create disease-resistant carp strains.

Research into the use of probiotics and immune stimulants has shown encouraging results in improving carp overall health and decreasing illness vulnerability.

By prioritizing proactive measures, the business may reduce its dependence on reactive treatments and ensure the long-term viability of carp farming.

2. Market Demand and Economic viability.

Meeting the increasing demand for carp products while maintaining economic viability is a tricky balancing. Overproduction and market saturation may cause price swings and financial difficulties for farmers. Sustainable solutions include market diversification, value addition via processing, and the creation of niche markets for specialist carp products.

Furthermore, certification organizations such as the Aquaculture Stewardship Council (ASC) assist customers in identifying and supporting sustainably grown carp.

Farmers that follow ASC guidelines not only contribute to environmental protection but also gain a competitive advantage in the market by satisfying rising consumer demand for sustainably produced seafood.

Success Stories Of Carp Farming

Several success stories demonstrate the beneficial effects of sustainable techniques in carp farming. Farms that have adopted ecologically friendly approaches, such as IMTA and RAS, claim higher water quality, lower environmental impact, and greater production efficiency.

One famous example is a carp farm in Southeast Asia that successfully deployed a closed-loop RAS system. By dramatically lowering water usage while maintaining ideal circumstances for carp development, the farm not only saved money but also helped to preserve local water supplies.

Another success story includes a carp producer in Europe who has embraced organic growing procedures. The farm gained organic certification by removing synthetic chemicals and establishing a natural, balanced ecology inside the ponds. This not only attracted ecologically aware customers but also resulted in higher pricing for sustainably farmed carp.

Conclusion

Carp aquaculture requires sustainable techniques to ensure its long-term survival. Addressing environmental impact, maximizing resource use, and overcoming hurdles with new solutions are critical steps toward a sustainable future for carp farming. The success stories of farms that have adopted these techniques demonstrate the excellent results that may be achieved via a dedication to sustainability.

By constantly improving and deploying ecologically friendly processes, the carp aquaculture business may help to ensure global food security while also protecting aquatic ecosystems.

As the need for fish protein grows, adopting and promoting sustainable techniques becomes not just a duty, but also a strategic benefit for the future of carp aquaculture.